Copyright: Published in the United States by Nona J. Fairfax / © Nona J. Fairfax

Published 02/02/2015

All rights reserved. No part of this publication may be reproduced, stored in retrieval system, copied in any form or by any means, electronic, mechanical, photocopying, recording or otherwise transmitted without written permission from the publisher. Please do not participate in or encourage piracy of this material in any way. You must not circulate this book in any format. Nona J. Fairfax does not control or direct users' actions and is not responsible for the information or content shared, harm and/or actions of the book readers.

# The Blue Mermaid and the Little Dolphin Book 2

HELP TRAIN YOUR CHILDREN'S IMAGINATION

# Introduction

For many parents, curling up with a book for a bedtime story with their kid is a daily ritual. For others, it is the perfect time to spend time with their children after a busy day, and for some, it is something they should do but are not entirely sure why. Discover these benefits of bedtime stories for kids.

## Sharpen their brains

Research shows that one of the greatest benefit of interacting with children, including reading to them stories, is that children learn a great deal of things- from improved logic skills to lowering their stress levels. Bedtime stories rewire the brain of a child and quicken their mastery of language. Their vocabulary repertoire is expanded and their listening and oral communication skills enhanced.

## Enhance creativity and Stimulate imagination

If you are a good storyteller, then you should teleport your kid to a different realm- from reality to fantasy for the child to learn the difference between these two. This will enhance and stimulate his imagination.

## Emotion development

The kid will learn to experience different emotions while empathizing with the characters of the story. The common emotions of sadness, happiness and anger may be encountered and he will learn to control these in real life.

## Content

- Introduction
- 7 Tips to Help Ignite Your Child's Imagination
- The Blue Mermaid and the little dolphin Book 2
- Check Out My Other Books

# 7 Tips to Help Ignite Your Child's Imagination

Imagination, often the concept we associate with joy, whimsy, and play, is just as easily involving cognitive capability, logical reasoning, and mental stability. The process the brain goes through when utilizing make-believe notions is a mixture of curiosity, adrenaline, and the practicing of creativity and exploration of different ideas previously unreached. Doing so exercises several parts of the adolescent mind, and will be reflected upon in the future, by means of open-mindedness, reason, and the ability to fathom abstractions above and beyond that of the average mind.

By incorporating these tips and activities into the lifestyles of both you and your child, your pudgy genius will not only be more mentally efficient, but happier and more capable overall.

1. Ask Boggling Questions

Parents tend to underestimate the span of their child's intelligence, restraining their capacity. Ask questions that are "out of the box", for instance, "what do you think that toad is thinking?" or "how come the clouds are floating?" Questions that you wouldn't normally consider completely sane, probably absurd to the outside world, can be awe-inspiring and full of wonder to your child, opening doors of imagination and invention in their minds. In addition, try to respond to their answers with "why do you think that is?" or "how can you tell?", thereby teaching them that questions are meant to be explored, and ideas full of possibility. Most importantly, never allow your child to say "I don't know". Teach them that your questions have no right answer, as long as they make you think.

2. Keep Questions Open-ended

Steer clear of words such as "wrong" or "try again", which will force children to believe that questions can only have one answer, with strict guidelines and a sole truth. They will later apply this mantra to their everyday behavior, and think in a more linear format, essentially restricting their minds and ideas.

3. Ask "what if?"
What if trees could move? What if you your teacher were an ogre? What if giraffes could talk? Let your child explore things beyond the dimensions of reality, and allow them to conceive concepts far and wide.

4. Tell Stories
Pictures and imagery are unimportant, even less effective than simply speaking a story to your child. Use different voices, adjectives, and hand gestures when speaking, allowing your child to fill in the blanks of the story, creating a sort of movie in their own heads. Unlike in TV, in stories, children do not have their visuals presented to them. Make these epic narratives engaging, and allow your child to add to the story. This will open the door to creative writing, as well as reading and extensive thought. It is so important to constantly exercise your child's mind, allowing it to be flexible and thrive in mentally straining instances in adulthood.

5. Create Art

Allow your child to openly explore their artistic talent. By letting your child draw, paint, or simply create the ideas in their minds, you are solidifying the concepts they invent, giving them meaning and allowing them to come to life. Even if their drawing capabilities are short of Mozart material, by creating art, your child will learn to express his or herself, perhaps opening up new areas of interest.

You can create hand puppets, form houses out of old packing boxes, or have fun with finger paints. The expanse of art is endlessly vast, and it's fine simply to explore its plains.

6. Encourage Openness

In addition to openness regarding the imagination, encourage trying new things. Whether this is varying foods, different school clubs, and even new friendships, your child can experience anything from an engaging hobby to understanding the conditions of friendship. Moreover, try not to overly shelter your child. Children who understand death at an early age will better cope with it in the future, and adolescents familiar with financial instability will learn from their parent's experiences and be more understanding and compassionate toward adversity. Of course, there is a period in childhood when not to be too open, just keep in mind that hiding everything from your child is not realistic nor healthy.

7.  Leave Room for Growth

A flower doesn't bloom under watch, and a child's imagination cannot grow in a cramped environment. Allow him or her space to explore the world. Let your child establish independence, while still maintaining the assurance of an adult figure, in case things go astray. It's often hard to let go of your child, and that's OK, however it's imperative to refrain from being overly invasive, as this will likely cause your child to become dependent and create a barricade between them and their full capacity.

# The Blue Mermaid and the little dolphin

**S**ky, Princess Alice, Jacob, and his dolphin, Roller, raced over the waves on the great ocean. Alice and Sky made many good friends after the Annual Deep Marine Festival. Among them all, Jacob and Roller were their favorites. They raced and explored the vast sea. They went to places no mermaid had been before, discovering new reefs and caves. Today they planned to hit the wind and ride over the waves.

"Faster, Sky… faster!" said Princes Alice, riding on Sky. Jacob and Roller were ahead of them. Roller stopped with a jerk, and so did Sky. The riders patted on their dolphins, but they were not ready to go ahead.

"What is it, Sky?" asked Princess Alice. "Shhuussshh Alice…. I think there are pirates there," said Jacob.

"Pirates? What are they?" asked Alice, very slowly, keeping her voice down.

"They are thieves of the sea. I wonder what are they looking for?" replied Jacob.

Jacob patted Roller, and they went below the surface, deep beneath the pirate ship. Alice and Sky followed them. It was a very dirty ship. It smelled very badly, even from far away. Jacob signaled Alice to stay, as he went closer to the ship to see what the pirates were up to. He went close to the ship to hear them. They were talking about catching mermaids. Jacob was shocked by what he had just heard… he swam back to Alice. But Alice was not to be found.

"Alice? Alice…… SKY?" he called out, louder each time he called their names.
Were Alice and the dolphins in trouble?
He went around the ship in search of them. As he swam below the ship, he felt entangled. Something was pulling him… Jacob swam as fast as he could, but he did not move. Rather, the ship got closer and closer, and Jacob found himself stuck in a large net that pulled him above the surface of the water.

He saw Alice in a large lank filled with water. The net was raised and shifted above the tank. He was thrown down. The water in the tank splashed out.

"Are you all right?" asked Alice.

"Yes are you? What happened how they found us?" the pirates stared at Jacob as he spoke. They did not understand the mermaid language but they were looking at their find and amazing that the mermaids communicated with each outer.

The captain of the ship walked up to them. "Mighty ho! wWe caught two of them. Isn't this our lucky day!" said the ugly captain and he spat.

"Close it boys... let's find more," and as he said that, two other pirates dropped down a large wired lid over the tank, covering the tank completely, so there was no chance of escape. Sky and Roller were free, but they swam next to the ship. Sky jumped up high.

Alice looked at her dolphin "Oh, dear Sky." She wanted to cry.

Jacob consoled her. But Alice was not ready to give up so soon. She knew she had to get out of there. Princess Alice looked around and built up a plan to get them out of the smelly, ugly pirate ship.

On the ship, Princess Alice observed the things that were kept on the deck. There were barrels stacked up. She didn't know what was in them, but she had to figure out how to use them some way. There were ropes hanging from the sails. They were of different sizes. She searched for the longest rope there. She found one rolled up at the edge of the deck. Next she observed the pirates. There were tall and skinny pirates, short and stout pirates, and there were bald and grumpy ones, too. They worked around the ship, but more than anything they fought among themselves. A tall pirate pushed a bald and grumpy pirate, hence starting a battle among the pirates.

The captain of the ship fired a rusty rifle, and the bullet hit one of the posts. It broke the rope tied on it into two. Its rope was holding some barrels that came down, crashing onto the fighting pirates.

They all fell down at once.
"Get back to work, you quarreling, lazy crooks," said the captain of the ship.

Jacob giggled, looking at the commotion on the deck. But Alice was more interested in plotting an escape.

"Jacob, we need to get out of here," she said. "I have a plan."
She hoped that Sky and Roller were still there near the pirate ship.

"What do you have in mind, Alice?"asked Jacob.

"We need to push this lid and then drop those barrels just like it happened now. Then when they all get busy clearing it up, we'll jump out of here," said Alice.

"It does not sound like an easy plan, Alice. It's too dangerous," replied Jacob.

"Yes, but there is no other way. I hope Sky and Roller are still out there; we can swim deep and hide." It was true there was no other way out. Jacob agreed with her, and even though he did not know how it would turn out, he decided that he had to try.

They planned that Alice would keep a look out while Jacob pushed the wired lid up slowly. It was heavy and his wet hands were slippery. He pushed it, inch by inch, until there was a wide opening. Lucky for them, no pirate observed them. Next was to drop the barrels above them. Jacob knew they needed help. Roller could help them. Growing up with Roller, Jacob had learned how to whistle like a dolphin. He whistled loud. Roller heard the whistles under the water. Roller and Sky did not leave their riders behind. Roller replied back by whistling.

"Yes, yes, they are here!" Jacob jumped with happiness.

"Shushhhh! We don't want to attract attention now, Jacob" said Princess Alice.

Jacob once again whistled, this time instructing Roller about what had to be done. The distance was long. Alice prayed for her plan to work, and mainly for no one to get hurt.

Jacob was confident about Roller; Roller had won many awards for the high jump and diving. It was noon time. The captain had halted the ship and walked into his cabin for lunch, and the pirates on the ship were busy with some chores and dousing in the noon heat. Meanwhile, Alice took the best of the opportunity and tried to drag the rope near her tank closer to her. It was difficult, as her paws were wet and slimy, like all marine creatures. Alice grabbed on to a hook and threw it over the ropes. After several failed attempts, the rope finally came within her reach. She noticed none of the pirates were looking over at them.

"Now is the right time," said the princess.

Jacob gave one loud whistle, and Roller was ready. He swam deep and swam back up again at full speed. He launched his body out of the water and onto the deck. He pulled the ropes, untying the barrels, and dove back into the ocean. It was a perfect jump. Jacob leaned over Alice and covered her in his arms to protect her. The barrels broke loose and dropped down, causing banging noises on the deck. The pirates came running because they didn't know what had caused the barrels to fall. They were worried it would upset the captain. Hence they all got into clearing the mess. Alice tried to throw the rope to one of the lowest sails to her reach. Jacob took it from her and tried. He was successful on the first attempt.

"Now call them," commanded Alice.
Jacob once more whistled loudly. Roller once again dove onto the deck – this time, instead of breaking the rope, he caught it with his strong jaws.

"Hold on tight... this won't be easy," said Jacob. Jacob grabbed one end of the rope and Alice grabbed the other end. The top of the rope, which was hooked by Roller's jaws, lifted both Alice and Jacob out of the tank that they had been kept captive in. Before the pirates even turned their heads, Roller dove back into the ocean, taking the captive mermaids along with him.

The captain was alarmed and came out running. He shot bullets into the ocean, but it was of no use. Alice rode Sky and Jacob rode Roller and escaped as far as they could. But the captain was not ready to let go of his prisoners so easily. He pulled out his telescope and watched the sea. Not very far away, he saw the mermaids riding the dolphins.
"Ready the host," he said. "Follow them!" He pointed his telescope towards the direction they were heading.

Back in the ocean, the dolphins swam as fast as they could.
"Don't you think we should head home, Princess? It will take us hours to swim back home if we go far away," said Jacob

"Trust me. This is the safest way to reach home," said Alice. She knew the captain was a stubborn man; he would hunt them down, so going back home now would be inviting the pirates into their kingdom. This was all the part of her plan, and the captain proved her right by following them and raising an attack on them.

"We've got to do this. We can trap their ship in the creek," said Alice. It was a solid plan, but a very risky one. No marine animal went to the creek because the currents would drag them into the narrow walls of the creek and drown them in it, with no chance of escape. It was also a dangerous place for any ship sailing in the area. The currents would drag the ship into the walls of the creek, and once a ship was trapped, the currents would tear it, bringing it down. There was no chance of escape.

"They will destroy us all. What if they discover our homes, Jacob? We cannot let our families be destroyed." She was right – if the pirates found out about the other mermaids, they would make them all captives. Pirates were heartless in such matters, especially the captain of this ship.

"You're right, but we need to think about it before going there," said Jacob.

Lucky for Arial, she remembered something from the ship that would help them out. The long ropes that helped them escape were still with her. She showed them to Jacob. He nodded, agreeing to her plan. Now all they had to do was lead the ship to the creek, and the currents would do the rest for them. They slowed down and jumped above the surface of the water to get the pirates' attention. The captain was too greedy, so he did as Alice had anticipated. He followed them. Faster and faster the ship sailed on the ocean. They were still aiming cannons at them, but none of them reached them. Alice, Sky, Jacob, and Roller dove deep. The creek was closing in; they could feel it by the strong currents pulling them. They quickly found a strong rock. Jacob took the rope from Alice and tied it to it. He made sure the knot was strong – then he tied Sky and Roller to the ends of the ropes. Princess Alice and Jacob now swam against the currents and rose to the surface. The pirates were waiting for them; looking through his telescope, he spotted them.

He did not bother to check where he was heading; his greed and pride, and desire to catch the mermaids, were too strong. He ordered his pirates to take the ship in the direction he had spotted the mermaids.
"Alice, please go to Sky; I will continue from here," said Jacob, as he was very worried.
"No, I will not leave you alone, Jacob; this was my plan, and I will be there till the end of it," said Alice. They saw the ship closing in. They swan ahead, and they could feel themselves being pulled down.

They both used all of their strength and swam to hold their position. But the ship was closing in. They swam back to their dolphins and embraced them tightly. From below, they saw the ship above them. Slowly, the ship started sailing towards the creek, and before the pirates knew it, they were in a trap. The ship lost control and got stuck in between the walls of the creek. Its currents were so strong it sucked the ship in, breaking it into many pieces and destroying it.

Jacob unleashed Sky and Roller. The riders rode their dolphins and swam back to safety. Princess Alice and her friends smartly saved themselves and the marine world form the evil pirates.

THE END.

## ABOUT AUTHOR

## Nona J. Fairfax.

Nona J. Fairfax is an accomplished storybook author that has a strong passion for writing when it comes to improving the lives of children. She has a strong sense of pride for her work, and continues to thrive for a more universal world that keeps children and families in the foremost front of her mind. Her life's work includes writing children's books that help to improve the parent-child relationship, bedtime stories, and smaller pieces of art. When she is not busy writing about improving the lives of children.

**MY BEST SELLER ON AMAZON**

Download on:
www.amazon.com/dp/B00YL3PJSK

Download on:
www.amazon.com/dp/ B00Y5DC4MU

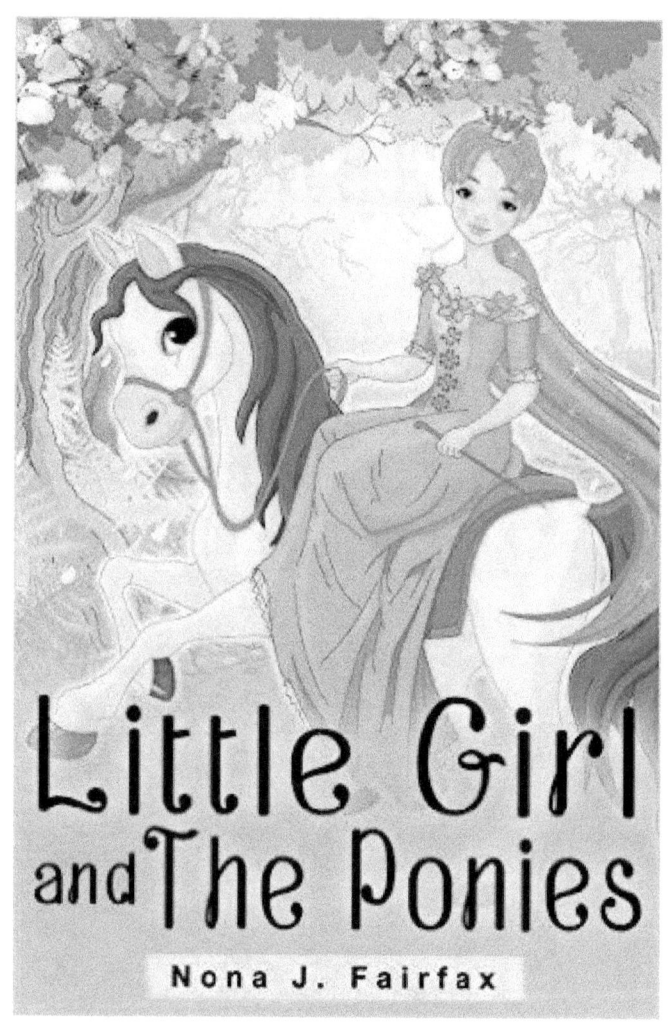

Download on:
www.amazon.com/dp/B00S2V6Z1W

Download on:
www.amazon.com/dp/ B010EAB0JA

Download on:
www.amazon.com/dp/B010LP0TJU

Download on:
www.amazon.com/dp/ B00WXI5HDM

www.ingramcontent.com/pod-product-compliance
Lightning Source LLC
Chambersburg PA
CBHW070430190526
45169CB00003B/1494